FLOATING
JELLYFISH

by Kathleen Martin-James

Lerner Publications Company • Minneapolis

This book is available in two editions:
Library binding by Carolrhoda Books, Inc., a division of Lerner Publishing Group
Soft cover by First Avenue Editions, an imprint of Lerner Publishing Group
241 First Avenue North
Minneapolis, MN 55401 U.S.A.

Website address: www.lernerbooks.com

Words in *italic type* are explained in a glossary on page 30.

Library of Congress Cataloging-in-Publication Data

Martin-James, Kathleen
     Floating jellyfish / by Kathleen Martin-James.
        p.   cm. — (Pull ahead books)
     Includes index.
     Summary: Introduces the physical characteristics,
habits, and natural environment of the jellyfish.
   ISBN 0-8225-3766-4 (lib. bdg. : alk. paper)
   ISBN 0-8225-3769-9 (pbk. : alk. paper)
   1. Jellyfishes—Juvenile literature. [1. Jellyfishes.]—
I. Title. II. Series.
QL377 .S4 M27 2001
593.5'3—dc21   00-008877

Manufactured in the United States of America
1 2 3 4 5 6 – JR – 06 05 04 03 02 01

This floating animal is a jellyfish.
Do you think it looks like a fish?

A jellyfish is not a fish. A jellyfish is shaped more like a mushroom.

Most jellyfish live in oceans. But some jellyfish live in lakes.

Jellyfish can be small like this pink and white one.

This jellyfish is about the size of your hand.

Jellyfish can also be very big.
But they start out very small.

You can see the eggs inside this adult jellyfish.

Each egg
becomes a
larva.

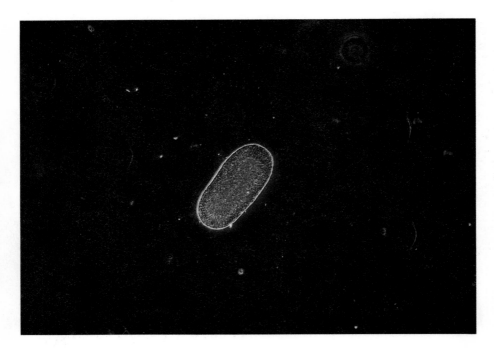

A larva looks like a floating
jellybean.

The larva attaches itself to something hard, like a rock.

Then it is called a *polyp.*

A polyp
clones itself
many times.

Cloning means it makes a copy
of itself. The copies stack on top of
each other like pancakes.

The copies break off one by one.
Each one becomes a jellyfish.

The new jellyfish grow quickly.

This jellyfish can glow in the dark.

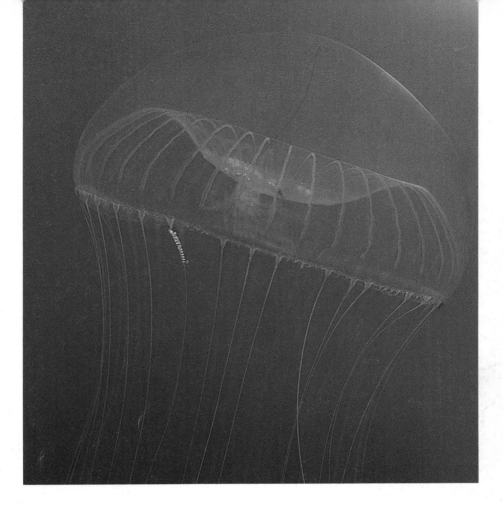

It is hard to see this jellyfish.
It has clear skin.

Look carefully at this jellyfish.
Can you see the *jelly* under its skin?

A jellyfish has no bones.
But it has muscles.

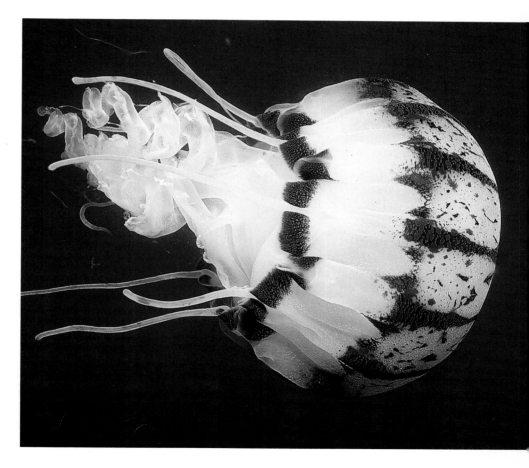

A jellyfish uses water to hold itself up.

A jellyfish pushes against the water to move up and down.

What happens when a jellyfish is washed onto land by waves?

It dries up. A jellyfish needs to be in water to live.

To swim, a jellyfish opens its body,
like an umbrella opening.

Then the jellyfish quickly squeezes
its body, like an umbrella closing.

Jellyfish also float in the water.

They have *arms* and *tentacles* that hang below them as they float.

The arms come out of the belly of the jellyfish. They are frilly.

Tentacles hang from the edges of a
jellyfish. They look like strands of hair.

Tentacles can sting.

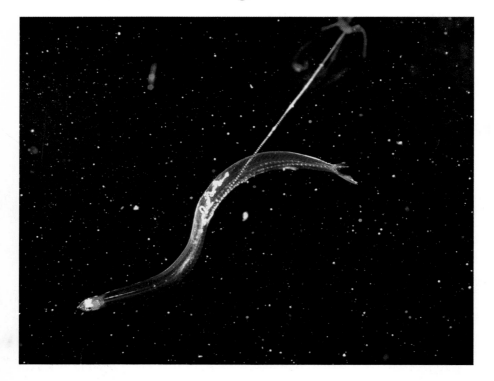

When a tiny fish or other animal gets
caught in the tentacles, it is stung.
It cannot move.

The jellyfish rolls the animal up to its belly and eats it.

Some animals are not hurt by the tentacles of a jellyfish.

What are these fish doing?

They are hiding from *predators*.

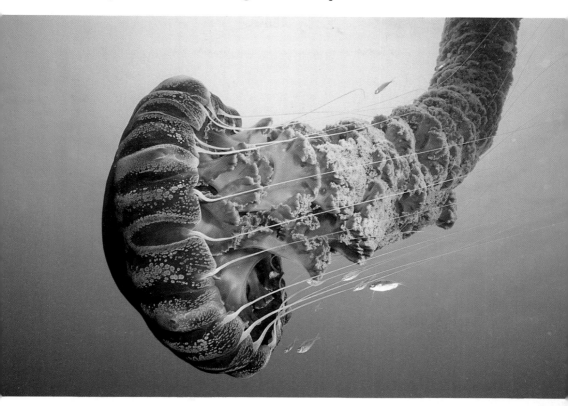

Predators are animals that hunt and eat other animals.

Jellyfish are hunted by predators too.
Many sea turtles eat jellyfish.

These floating jellyfish are safe
for now.

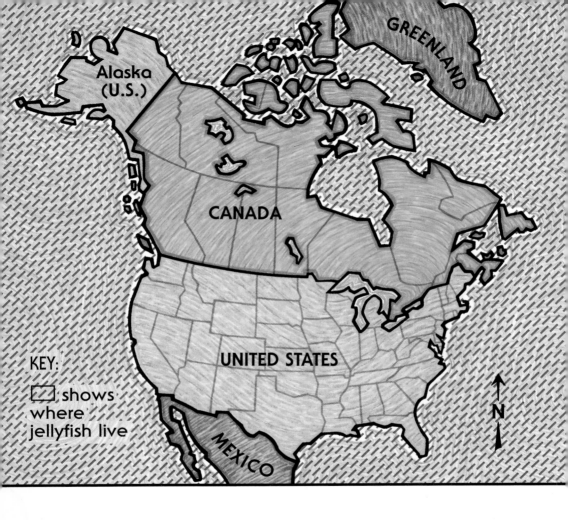

Find your state or province on this map.
Do jellyfish live near you?

Parts of a Jellyfish's Body

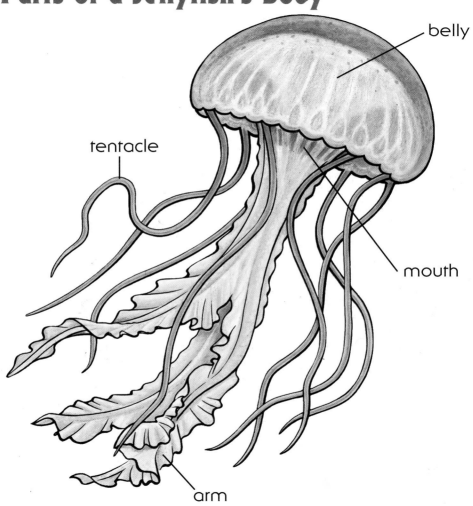

belly

tentacle

mouth

arm

29

Glossary

arms: wavy bands that hang down from the mouth of a jellyfish

clones: makes an exact copy of an animal

jelly: the substance found inside a jellyfish, between the layers of skin

larva: the form of a jellyfish that looks like a jellybean

polyp: the form of a jellyfish that has attached itself to a rock or other hard object

predators: animals that hunt and eat other animals

tentacles: hairlike strands that hang from the body of a jellyfish and sting

Hunt and Find

- a polyp **cloning** itself on page 10
- a jellyfish **larva** on page 8
- a jellyfish **glowing** in the dark on pages 4, 5, 12, and 19
- a **dried-up** jellyfish on page 17
- fish **hiding** near a jellyfish on pages 24 and 25
- long **tentacles** of a jellyfish on pages 4, 13, 15, 16, 18, 19, 20, 21, 24, and 25

The publisher wishes to extend special thanks to our **series consultant,** Sharyn Fenwick. An elementary science-math specialist, Mrs. Fenwick was the recipient of the National Science Teachers Association 1991 Distinguished Teaching Award. In 1992, representing the state of Minnesota at the elementary level, she received the Presidential Award for Excellence in Math and Science Teaching.

About the Author

Mike Dembeck

Kathleen Martin-James was born in Toronto, Ontario. She has lived in many different places across Canada and in the United States. In the summer, she and her husband, Mike, have many chances to watch jellyfish float in the Atlantic Ocean near their home in Halifax, Nova Scotia.

Photo Acknowledgments

The photographs in this book are reproduced through the courtesy of: © Ken Lucas/Visuals Unlimited, front cover, p. 16; © James McCullagh/ Visuals Unlimited, back cover, p. 12; © Richard Herrmann/Innerspace Visions, p. 3; © Gregory Ochocki/Innerspace Visions, pp. 4, 13; © Dave B. Fleetham/Tom Stack & Associates, p. 5; © Bob Cranston, pp. 6, 20, 21, 24, 25; © David Wrobel/Innerspace Visions, pp. 7, 14; © Tom Stack/Tom Stack & Associates, p. 8; © Ben Cropp/www.norbertwu.com, p. 9; © Cabisco/Visuals Unlimited, p. 10; © Triarch/Visuals Unlimited, p. 11; © Randy Morse/Tom Stack & Associates, p. 15, 31; © John D. Cunningham/Visuals Unlimited, pp. 17; © Mark Strickland/Innerspace Visions, pp. 18, 19; © Doug Perrine/Innerspace Visions, pp. 22, 26, 27; © Fred Bavendam, p. 23.